ENERGY FOR THE FUTURE AND GLOBAL WARMING: WIND POWER

By
Nigel Saunders

16
EasyRead Large

RHYW

Copyright Page from the Original Book

Please visit our web site at: www.garethstevens.com
For a free color catalog describing Gareth Stevens Publishing's list of high-quality books, call1-800-542-2595 (USA) or 1-800-387-3178 (Canada).

Library of Congress Cataloging-in-Publication Data

Saunders, Nigel.
Wind power / Nigel Saunders.
p. cm. — (Energy for the future and global warming)
Includes index.
ISBN: 978-0-8368-8405-0 (lib. bdg.)
ISBN: 978-0-8368-8414-2 (softcover)
1. Wind power—Juvenile literature. I. Title.
TJ820.S28 2008
621.4'5—dc22 2007008755

This edition first published in 2008 by
Gareth Stevens Publishing
A Weekly Reader® Company
1 Reader's Digest Road
Pleasantville, NY 10570-7000 USA

Produced by Discovery Books
Editors: Geoff Barker and Sabrina Crewe
Designer: Keith Williams
Photo researcher: Rachel Tisdale
Illustrations: Stefan Chabluk

Gareth Stevens editor: Carol Ryback
Gareth Stevens art direction and design: Tammy West
Gareth Stevens production: Jessica Yanke

Photo credits: PPM Energy: / cover, title page, 17, 20, 22. NASA: / 6. istockphoto.com: / 8, 15, 21. istockphoto.com: / Don Bayley 10. Library of Congress: 13. Horizon Wind Energy: / TTweak/Horizon / Maple Ridge Wind Farm: 25. Bergey Windpower: / 26. Ben Shepard: / Sky Wind Power Corporation / Professor Roberts 28.

Printed in the United States of America

1 2 3 4 5 6 7 8 9 11 10 09 08 07

Cover photo: Maple Ridge Wind Farm in Tug Hill, New York, is the largest wind farm in the state. Construction began in May 2005. The first wind turbine began generating power in December 2005. All 120 wind turbines were completed by the beginning of 2007.

This optimized ReadHowYouWant edition contains the complete, unabridged text of the original publisher's edition. Other aspects of the book may vary from the original edition.

Published in 2008 in the US and world markets by ReadHowYouWant.

Set in 16 pt. Verdana

TABLE OF CONTENTS

CHAPTER ONE

ENERGY AND GLOBAL WARMING

Electricity is a very useful type of energy. Many things you do are helped in some way by electricity.

The demand for energy

Worldwide, people use huge amounts of fuel to make electricity. They use fuel for vehicles, too. The world's population is growing fast. As each country's population grows, the demand for fuel and electricity increases. At the same time, more countries are developing. They are building new industries and buying more goods. Developing nations are demanding more energy.

Where does the energy come from? Most of the world's energy comes from coal, oil, and natural gas. These fuels are **fossil fuels.** Fossil fuels formed in the ground over millions of years. They formed from the remains of plants and animals. About half of the electricity used in the United States is made from coal. Our vehicle fuel comes from petroleum, or oil. We use natural gas to heat many of our homes and businesses.

Burning fossil fuels

Fossil fuels make a lot of **pollution** when they burn. The burning fuel gives off smoke that goes into the air. Burning fossil fuels also give off harmful gases. Sulfur dioxide is a gas that causes acid rain. The term "acid rain" includes acid rain, sleet, and snow. Acid rain kills plants and even entire forests. (Image 1.1)

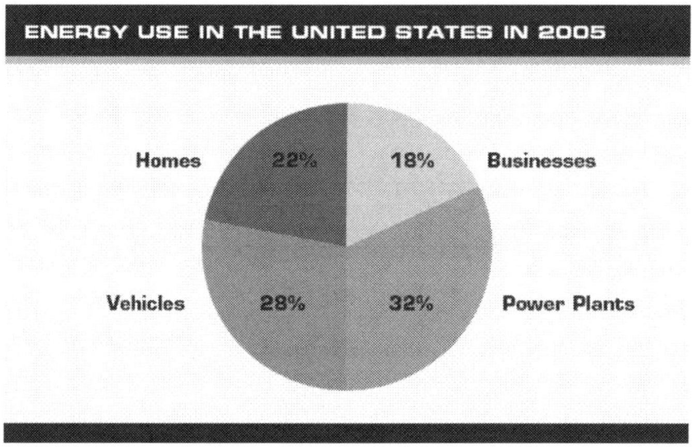

Image 1.1: This chart shows energy use in the United States. It shows how much was used by homes, businesses, power plants, and vehicles.

Greenhouse gases

Burning fossil fuels also give off **greenhouse gases,** such as carbon dioxide. Water vapor and methane gas are greenhouse gases, too. Greenhouse gases trap heat in the atmosphere. They keep Earth warm enough for life. When Earth's atmosphere traps too much heat, it becomes warmer than usual.

The amounts of greenhouse gases in the air have increased in the last one hundred years. Scientists believe this increase is warming Earth. Higher temperatures change global weather patterns, called the climate. This climate change is called **global warming.**

Global warming

Global warming is affecting the whole planet. At the North and South Poles, global warming is causing glaciers and large sheets of ice to melt. Water from the melted ice flows into the sea and causes sea levels to rise. As the sea levels rise, coastal lands will become flooded.

THE AYLES ICE SHELF

The Ayles Ice Shelf was in northern Canada. For hundreds of years, winds blew sea ice onto the northwest edge of Ellesmere Island. The ice formed a large, thick ice plate that looked like a shelf. It covered about 25 square miles (66 square kilometers). In August 2005, this huge slab of ice suddenly broke away from the island. Scientists believe that warmer Arctic temperatures and strong winds caused the split. They estimate that 90 percent of Ellesmere Island's ice shelves have disappeared since 1900. (Image 1.2)

Image 1.2: A satellite image shows Ellesmere Island. The Ayles Ice Shelf (red circle) has broken away from the island and is moving out toward the ocean.

While some areas of the world are too wet, others are too dry. As global warming increases, certain areas will get too little rain. Grasslands or forests may become deserts. In some farming regions, the climate may become too hot and dry for growing crops.

Renewable resources

We need to do all we can to reduce global warming and pollution. To achieve this, we must use fewer fossil fuels. Another reason to start using other energy sources is that fossil fuels cannot be replaced once we use them up. We must start using more of the other sources of energy.

Renewable energy sources will never run out. They also make few harmful gases. One type of renewable energy we can use is wind energy.

Using wind energy

In the past, wind was used to drive machinery. Today, we use wind power to make electricity. Only about 1.25 percent of the world's electricity is generated this way. In the United States, less than 0.5 percent of electricity comes from wind power. This will soon change. The U.S. Department of Energy has set a goal. By 2020, the United States will make 5 percent of its electricity from wind.

KEY WORDS

global warming: the gradual warming of Earth's climate

greenhouse gases: gases in the atmosphere that trap heat energy

pollution: making land, air, or water dirty

renewable: having a new or reusable supply of material constantly available

CHAPTER TWO

A WINDY WORLD

If you ever rode your bicycle against the wind, you know how powerful the wind felt. Wind has a lot of **kinetic energy**—the energy of motion, or movement. The faster the wind blows, the more kinetic energy it has. The wind's force makes it harder for you to pedal your bike against it.

Where winds come from

Why does the wind blow? Heat energy from the Sun heats the atmosphere. This warming happens over water or land. As air warms, it expands, or spreads out. The warmed air becomes less dense (thick and heavy) and rises. Colder, denser air moves in to replace the rising air. We feel this movement of air as wind. (Image 2.1)

Image 2.1: A wind sock is a low-tech indicator of wind direction. A strong wind fills the wind sock and makes it fly almost parallel to the ground.

POWERFUL WINDS

Some winds move at very high speeds. Earth's highest recorded surface wind speed was 231 miles (372 kilometers) per hour. It occurred during a wind gust on Mount Washington, New Hampshire, on April 12, 1934. Some winds cause great damage. Tornadoes occur when warm air and cold air collide. The air masses begin spinning, which pulls in more air from the ground. This forms a spinning funnel of wind. One 1999 tornado in Oklahoma had a record speed of 318 miles (512km) per hour. When tornadoes touch the ground, they can lift up cars and move houses. Hurricanes happen when warm air rises above the ocean. As it rises, the air begins to spin. It creates a storm of high winds, huge waves, and rain. Hurricanes are the largest storms on Earth.

Earth spins around once every twenty-four hours. This movement makes the winds swirl in different directions. A wind is named after the direction from which it blows. A westerly wind blows from the west. A wind blowing from the east is an easterly wind.

Onshore and offshore winds

Land heats up faster than bodies of water. As sunlight heats land, air above the land warms up. That warm air rises, and wind blowing from the water takes its place. Wind that blows onto land is called an onshore wind.

The opposite happens at night. Water keeps its heat longer than land. Warm air above the water rises. Cooler air blows from the land toward the water. This is called an offshore wind.

Some places are much windier than others. Large flat areas, such as plains and oceans, tend to have more wind. There are no trees or mountains to slow the wind down. (Image 2.2)

Image 2.2: A kite surfer gets an exciting ride thanks to the wind. The stronger the wind, the faster the surfer speeds through the waves. Cowabunga!

A working wind

Blowing wind provides free energy. Even more important, wind gives us clean energy. Wind power does not give off emissions (exhaust gases sent into the air) that add to global warming.

The wind does useful work by making objects move. **Windmills** are machines that use wind to power other machinery. A **wind turbine** is a kind of windmill. Its large, spinning blades make electricity.

Some sailboats use a variety of differently shaped sails to catch the wind. Before engines were invented, sailing ships carried people and goods across the oceans.

Sometimes, we use the wind for fun. Kite flying is an enjoyable pastime. As the wind pushes against the kite's sail, it soars into the air. Larger sails are useful for windsurfing, sailboarding, and kite surfing. Wind blowing into a sail pulls each craft through the water.

Whether for work or fun, people have found many ways to use wind power.

KEY WORDS

emissions: waste substances discharged into the air, land, or water

kinetic energy: the energy of movement

wind turbine: an engine with blades that produces electricity from the wind

windmill: a machine with blades that catches the wind and changes that kinetic energy into power usable for many purposes

CHAPTER THREE

CAPTURING WIND

Windmills have been in use for at least one thousand years. Large windmills were once common in many parts of Europe. They were used to grind wheat to make flour for bread. Windmills powered other machinery, too. They ran pumps used to drain wet ground. Draining made the land suitable for buildings and farms.

A sailing ship uses large sheets of material to catch the wind. Traditional windmills have sails, too. A windmill's sails, or blades, are made of wood. They are fixed to the top of a tower called the cap. Wind spins the sails around a **horizontal axis**—a shaft or support parallel to the ground. (Image 3.1)

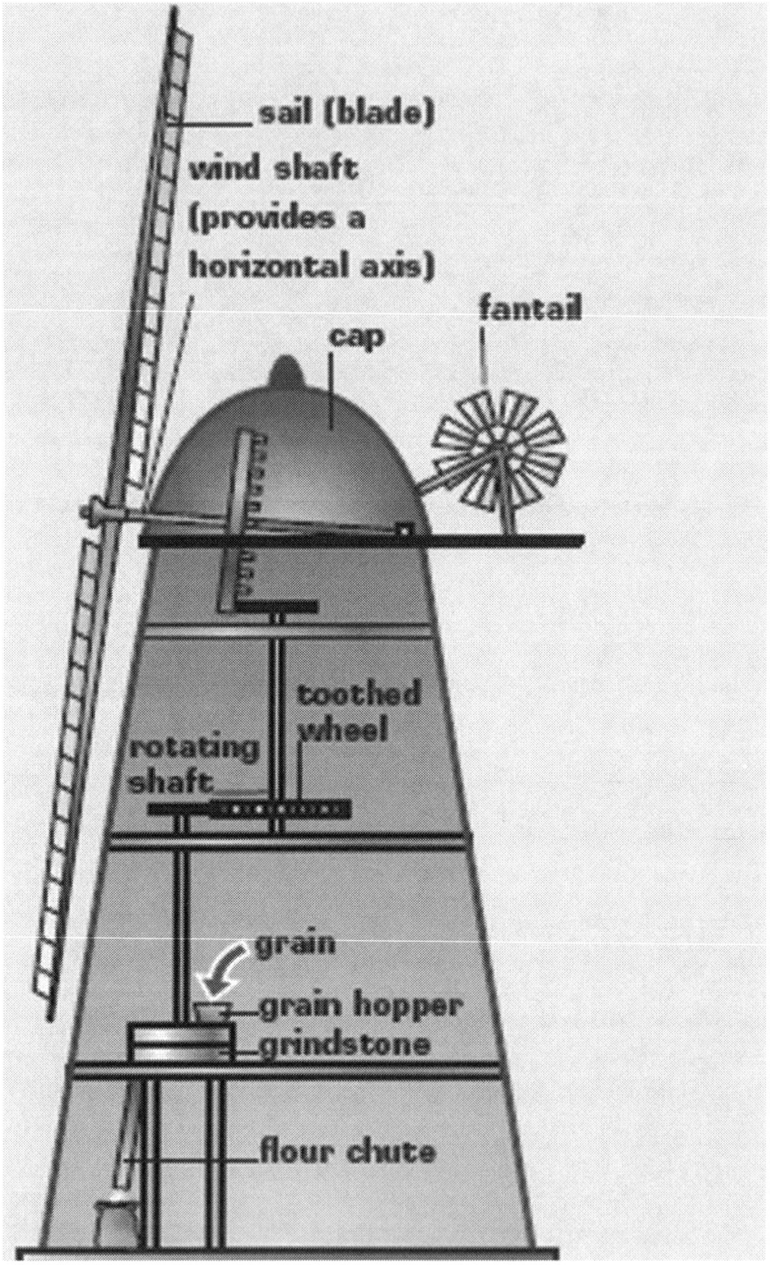

Image 3.1: This diagram shows the main parts of a traditional windmill.

A windmill has another, smaller set of sails called the **fantail.** The fantail is connected at a right angle to the main windmill blades. The fantail helps keep the windmill facing into the wind. If the wind changes direction, the fantail spins to turn the cap. That way, the main blades can always catch the strongest wind.

American farm windmills

Early U.S. farmers also used traditional windmills. The windmills were sometimes hard to run. They were often damaged by the wind. Daniel Halladay was a mechanic from Connecticut. In 1854, he invented a simple wooden windmill. It could control its own speed and turn to face the wind. (Image 3.2)

Image 3.2: Wooden windmills like this one in Nebraska were once very common on farms across the United States. This photograph was taken in 1886.

VENTILATORS

Delivery trucks can become very hot in summer. They need a **ventilator** to remove the hot air. A ventilator is a device that provides fresh air. Anton

Flettner from Germany invented a wind-powered ventilator in the 1920s. His design is still used today. Flettner's ventilator uses a Savonius turbine connected to a fan. This turbine spins when the vehicle moves forward or when the wind blows. Moving air turns the fan, which removes hot air. No electricity is needed, and there are few moving parts. Flettner's ventilator is cheap to make and easy to maintain.

Buildings also get hot in the summer. A roof ventilator removes hot air from the building to help keep it cool. The hot air rises and leaves the building through the ventilator. If the wind blows, the ventilator spins and draws the air out faster. Solar cells also power some roof ventilators.

Halladay's windmill was used to pump water from underground. This made it much easier for settlers to get water to use on their farms. Over time, the design was gradually improved. Metal windmills with many curved blades became popular. These were better at using the energy in the wind than the flat wooden blades. Windmills like this are still made and used throughout the United States.

Savonius windmills

Sigrid Savonius from Finland invented an unusual windmill in 1922. The Savonius turbine has two blades.

They spin around a **vertical axis** (an up-and-down shaft around which the blades turn). If you view a Savonius turbine from the top, you can see that the blades form an "S" shape. The blades always face the wind, whatever its direction. (Image 3.3)

Image 3.3: Metal windmills like this one are used on farms across the United States to pump water from deep underground.

Savonius turbines are not very **efficient** at using wind energy. They waste more energy than a regular windmill. But Savonius turbines are simple and cheap to make. They are sometimes used to make signs rotate. More often, Savonius turbines are used to run ventilators on vehicles.

KEY WORDS

fantail: a small set of sails that keeps the windmill turned into the wind

horizontal axis: a support parallel with Earth around which a device turns

ventilator: a machine or structure that aids in air movement

vertical axis: an up-and-down support around which a device turns

CHAPTER FOUR

ELECTRICITY FROM WIND

The wind's kinetic energy (energy of movement) can be used to make electricity. Windmills that do this are called wind turbines. A wind turbine powers a **generator**—a machine that changes the turbine's energy into electrical energy, or electricity.

A typical wind turbine produces enough electricity for four hundred homes. If fossil fuels were used to power that many homes, they would give off about 2,800 tons (2,540 tonnes) of carbon dioxide each year. Using wind energy helps reduce levels of greenhouse gases in the air. (Image 4.1)

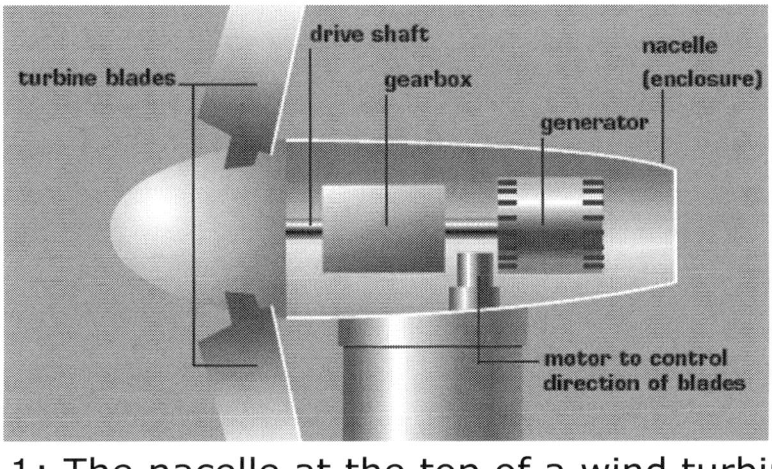

Image 4.1: The nacelle at the top of a wind turbine tower encloses the generator.

WIND POWER

GOOD THINGS	PROBLEMS
Renewable	Best in locations with steady winds
Free source of power, low running costs	Wind turbine construction costs often expensive
Leaves land under turbine available for farming	Wind turbines kill many birds
No waste or air pollution	Wind turbines spoil views; can create sound pollution
Good power source for remote areas	Best locations remote from large population centers

How wind turbines work

Winds blow stronger higher off the ground. Wind turbines are built with tall towers to catch more wind. Faster wind carries more energy. If the wind speed doubles, its energy increases eight times. Large blades catch more wind, so it can be economical to build a tall tower with large blades. (Image 4.2)

Image 4.2: The door at the base of this wind turbine allows maintenance personnel to enter for repairs.

STOPPING AND STARTING

Wind can be too weak or too strong for proper function of a wind turbine. The lowest wind speed needed to start a typical wind turbine is 9 miles (14.5km) per hour. This is called the cut-in speed. The average wind speed at which most wind turbines produce electricity is 13 miles (21km) per hour. Winds that blow faster than 55 miles (88.5km) per hour will damage most wind turbines. Wind turbines automatically shut off if the wind gets too strong.

The streamlined nacelle contains the generator and control equipment. The turbine's blades are fixed to the nacelle. Most turbines have three blades, but some have two. Both designs get about the same amount of energy out of the wind. Two-bladed turbines catch less wind, but spin faster than turbines with three blades.

The blades are slightly curved to catch as much wind energy as possible. The nacelle pivots (turns) so the blades always face into the wind. Each blade is usually about half as long as the height of the turbine. A turbine's height depends upon its location.

The turbine changes the kinetic energy of the blowing wind into mechanical energy. The mechanical energy of the spinning blades powers the generator. The generator changes mechanical energy into electrical energy. (Image 4.3)

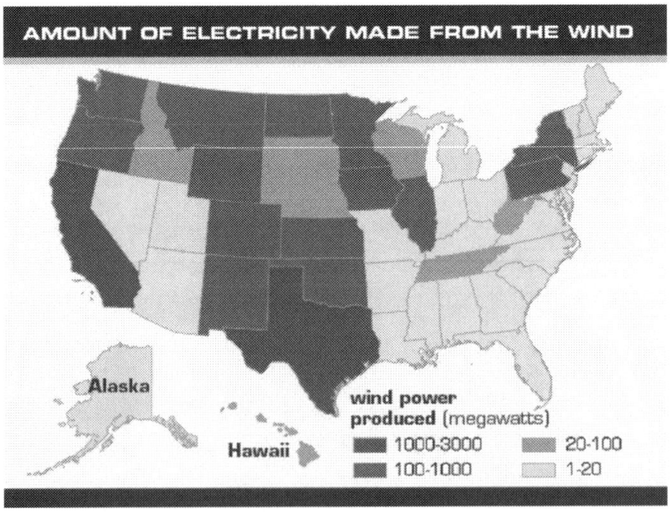

Image 4.3: Wind power is measured in megawatts. This map shows how much wind power was produced in each U.S. state during 2006.

Wind farms

Single wind turbines are often used to provide electricity for one farm. One turbine can also provide enough power for a few buildings in remote areas. To create more energy, wind turbines are linked together in groups called wind farms.

Electricity from a regular power plant travels through a network of cables. The network provides electric power to places that need it. Such a network is called an **electric power grid.** Wind farms can also be connected to the power grid. Power from the wind turbines travels through the same power lines used by regular power plants.

THE STATELINE WIND PROJECT

The Stateline Wind Project is a large wind farm on the border between Oregon and Washington states. Its first turbines were installed in 2001. By 2006, the project had 186 wind turbines. Eventually, the wind farm will have 465 wind turbines. It will be one of the largest in the world. Wind speeds at the project average 17 miles (27km) per hour. The turbines start turning at about half that speed. They automatically turn off if the wind gets faster than 58 miles (93km) per hour. (Image 4.4)

Image 4.4: Wide-open croplands are perfect for wind turbines. Farmers can continue to farm the land below the turbines.

Wind farms should be located where the wind is strong and steady. They are often located on hilltops or open plains. Other windy places are found in gaps between mountains and along coasts.

Designing a wind farm

Wind farms must be designed carefully. Wind turbines are big machines. People will complain if a wind farm spoils their view. Spinning blades can make annoying, flickering shadows on the ground. They also create noise. For these reasons, turbines cannot be too close to homes and other buildings.

The turbines must be spaced so that they do not block each other's wind. Wind farms can also affect bird life. Wind farms should not be built in birds' regular flight paths or in their seasonal, migratory routes. Still, hundreds of thousands of birds die from collisions with turbines every year.

Offshore wind farms

Image 4.5: Offshore wind farms are more expensive to build and maintain than those on land. Such turbines must be large enough to catch strong sea breezes.

Wind farms built along a coastline, where the wind is fast and steady, are called offshore wind farms. These turbines are anchored to the seabed. Offshore turbines may be more than 400 feet (120 meters) high. People don't complain as much about the size and noise of offshore turbines.

"Coastal wind power has come to the United States and found a home in Texas."

Jerry Patterson, Commissioner of the Texas General Land Office, speaking in 2005

Offshore wind farms are expensive to build. Wind turbines in the ocean must be protected against the salty air and water, or they will rust. Long, waterproof cables bring the electricity to shore. The cables must be laid deep on the ocean floor. Offshore wind farms can produce a lot of electricity with their large turbines. The power they produce makes up for the high cost of building them. The first U.S. offshore wind farm will be built 7 miles (11km) off the coast near Galveston, Texas. It should be completed in 2012.

Around the world

Wind power is an important energy source in Denmark. It manufactures about half of the world's wind turbines. Denmark gets about 20 percent of its electricity from wind power. Globally, Denmark ranks fifth in wind energy production. That may sound like a lot, but Denmark is a small country compared to most. (Image 4.6)

Image 4.6: Wind turbines must be positioned to make the best use of the main wind direction.

NATIVE WIND

Native Wind is a Native American organization. It promotes the use of wind power on tribal lands. Indian reservations in North and South Dakota are located in areas that make it possible to produce huge amounts of wind-generated electricity. This would benefit the local people. It would also reduce emissions of greenhouse gases. Native Wind plans to use its wind turbines for electricity and hopes to teach others about the benefits of using wind power on tribal lands.

Germany makes more electricity from wind than any other country. In 2006, Germany produced about one-third of the world's wind power. Other countries are catching up fast. Spain and the United States are building new wind turbine farms faster than Germany.

The United States produced about one-fifth of the world's wind power in 2006. More than thirty states have wind farms. California and Texas lead the nation in producing wind power *(see map, Image 4.3).*

Is there enough wind?

Wind turbines do not make electricity when the wind speed drops below a certain point. In most large states or countries, the wind blows fast enough somewhere for wind turbines to generate electricity constantly. Wind power could probably supply about 10 percent of almost any country's electricity needs. About 13 percent of the land in the United States is windy enough for wind turbines. Wind turbines could not replace every power plant. But they do make useful contributions to the electricity supply.

THE CAPE WIND PROJECT

People are often in favor of wind farms until one is built near their homes. For example, the Cape Wind Project in Massachusetts is planned for Nantucket Sound, south of Cape Cod. It will have 130 wind turbines. Nantucket Sound is a popular vacation spot. It is also a fishing ground. People who live and vacation there say the wind farm will spoil the view and harm fish. Supporters say using an energy source that may slow global warming is more important. They point out that global warming will harm ocean life much more than a wind farm.

It takes energy to build a machine as big as a wind turbine. After running for six months, a typical wind turbine makes up for the power used to build it. A wind turbine wears out after about twenty-five years. Then its parts can be recycled. (Image 4.7)

Image 4.7: Workers raise a rotor blade at Maple Ridge Wind Farm. Each blade is 130 feet (40m) long.

How they look and sound

Wind turbines are often painted a pale gray color to blend in more easily. Their electric cables run underground so they don't spoil views. Still, people worry that wind farms are unsightly. A wind turbine is not silent, either. The shape of its nacelle can reduce the amount of noise produced. Soundproofing also blocks the noise. The roar of the wind itself is usually louder than that of the turbine.

KEY WORDS

electric power grid: a network of cables and equipment that sends electricity from power plants to homes and other buildings

generator: a machine that turns mechanical energy into electrical energy

nacelle: the main body of a wind turbine that contains mechanical parts, such as the generator

CHAPTER FIVE

THE FUTURE OF WIND ENERGY

How will wind turbines and other wind power fit into our future? They will become more important. Engineers are developing new and better ways to capture the wind's energy.

Plans for the future

The U.S. Department of Energy has started a "Wind Powering America" program. It is asking energy companies to increase their use of wind power. Energy experts say about 13 percent of the U.S. is windy enough to power turbines to produce electricity. Cities in the Great Plains of the United States could save a lot of fossil fuels by using wind energy. (Image 5.1)

Image 5.1: Micro wind turbines like this one are becoming increasingly popular. They provide electricity for a single building without causing pollution.

NEW DESIGNS

Researchers at a company in Troy, New York, look for new ways to use renewable energy for buildings. The Wind Amplified Rotor Platform (WARP) is one of their ideas. It uses several small wind turbines to generate power. The turbines are placed on large, curved panels that catch the most possible wind. A WARP does not look much like a traditional wind turbine. (Image 5.2)

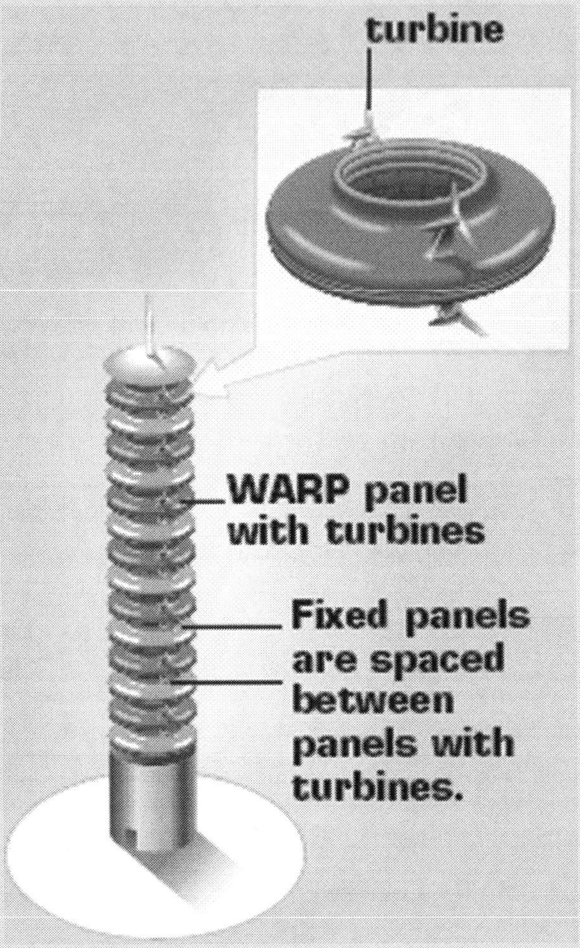

Image 5.2: The WARP design uses curved panels that channel the wind. The design boosts the wind speed through the turbines, which improves efficiency.

Bigger and smaller

The maximum amount of power produced by new wind turbines increases every year. Some of the biggest turbines have blades more than 200 feet (60m) long.

Engineers are always trying to make turbines more efficient. Some **experimental** wind turbines have a double set of blades to catch more wind energy.

Small wind turbines will also have a place in the future. Micro (smaller) wind turbines are designed to fit on private buildings. They provide electricity for just that structure. Micro turbines can provide clean power and save money for homeowners.

Turbines in the air

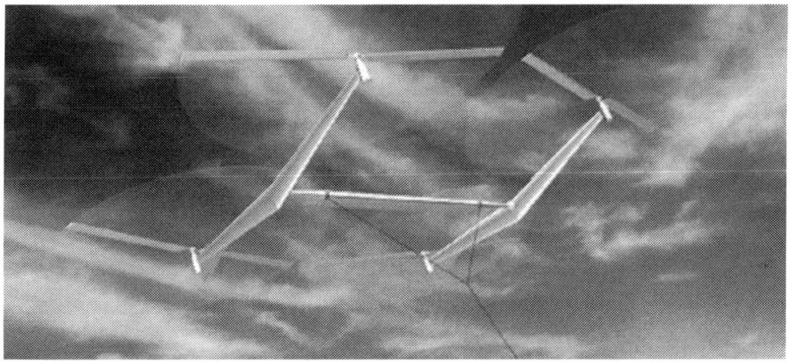

Image 5.3: Flying electric generators (FEGs) may someday be common in our skies. These wind-power devices must not interfere with air traffic routes.

Wind is much stronger high above the ground. One company is testing wind machines that generate electricity while flying in the sky! These devices are called flying electric generators (FEGs). They would attach to the ground by cables. FEGs would fly high at an altitude of about 9 miles (15km) above Earth. The spinning blades of the FEGs would keep them in the air. Electricity from the FEG's generator would reach the ground through the cables.

Storing energy for calm days

Wind energy is not the same every day. But we still need electricity on calm days. If we can find good ways to store wind power, it will be more widely used. Batteries can store extra energy from turbines on windy days. They can release it again on calm days.

"When I start off on a project I want to see its purpose ... I need to see that it could be important for something in the future. I also like to see a possible benefit to society."

Professor Maria Skyllas-Kazacos, inventor of the flow battery used at the King Island wind farm in Australia—from an interview with the Australian Academy of Science in 2000

A small wind farm on King Island, Australia, is using a battery called a flow battery. The flow battery pumps energy into a tank. The energy is stored in chemicals. As more tanks are added, more power can be stored. The flow battery used at King Island is an important development.

Future energy

No single energy source will replace fossil fuels. The world's demand for energy will be met by several sources. Clean sources of energy, such as wind energy, will play an important part in supplying our future energy needs.

KEY WORDS

battery: a cell that stores or provides electrical energy

experimental: having to do with an experiment (trying something new to see how it works)

micro wind turbine: a small wind turbine and generator for individual buildings

GLOSSARY

battery: a device that stores or provides electrical energy

efficient: working well and without much waste

electric power grid: a network of cables and equipment that sends electricity from power plants to homes and businesses

emissions: waste substances discharged into the air, land, or water

fossil fuels: fuels formed in the ground over millions of years from decayed plants and animals

generator: a machine that turns mechanical energy into electrical energy

greenhouse gases: gases in the atmosphere that trap heat energy

kinetic energy: the energy of movement

megawatt: a measurement of power produced. One megawatt is one million watts. One watt is the amount of electrical energy flowing in one second. Electrical energy is measured in joules. One watt is the same as one joule per second.

power plant: a factory that produces electricity

renewable: having a new or reusable supply of material constantly available

ventilator: a machine or structure that increases air movement in an area

wind turbine: an engine with blades that produces electricity from the wind

windmill: a machine that changes the wind's kinetic energy into usable power

TOP EIGHT ENERGY SOURCES

in alphabetical order

The following list highlights the major fuel sources of the twenty-first century. It also lists some advantages and disadvantages of each:

	Advantages	Disadvantages
Biofuels	renewable energy source; widely available from a number of sources, including farms, restaurants, and everyday garbage	fossil fuels often used to grow farm crops; requires special processing facilities that run on fossil fuels in order to produce usable biofuel
Fossil fuels: coal, oil, petroleum	used by functioning power plants worldwide; supports economies	limited supplies; emit greenhouse gases; produce toxic wastes; must often be transported long distances
Geothermal energy	nonpolluting; renewable; free source	available in localized areas; would require redesign of heating systems
Hydrogen (fuel cells)	most abundant element in the universe; nonpolluting	production uses up fossil fuels; storage presents safety issues
Nuclear energy	produces no greenhouse gases; produces a lot of energy from small amounts of fuel	solid wastes remain dangerous for centuries; limited life span of power plants
Solar power	renewable; produces no pollutants; free source	weather and climate dependent; solar cells expensive to manufacture

	Advantages	Disadvantages
Water power	renewable resource; generally requires no additional fuel	requires flowing water, waves, or tides; can interfere with view; dams may destroy large natural areas and disrupt human settlements
Wind power	renewable; nonpolluting; free source	depends on weather patterns; depends on location; endangers bird populations

RESOURCES

Books

Green, Jen. *Saving Energy.*
Improving Our Environment (series).
Gareth Stevens Publishing (2005)

Naff, Clay Farris (editor). *Wind.*
Fueling the Future (series)
Greenhaven Press (2006)

Parker, Steve. *Wind Power.*
Science Files: Energy (series).
Gareth Stevens Publishing (2004)

Web Sites

www.powerhousekids.com/stellent2/groups/public/documents/pub/phk_ee_re_001502.hcsp
Explore Alliant Energy's Kids' pages to learn more about wind power and other energy resources and download an activity book.

www.eia.doe.gov/kids/energyfacts/sources/renewable/wind.html
Visit the U.S. government's wind energy Web site geared to kids.

Publisher's note to educators and parents: Our editors have carefully reviewed these Web sites to ensure that they are suitable for children. Many Web sites change frequently, however, and we cannot guarantee that a site's future contents will continue to meet our high standards of quality and educational value. Be advised that children should be closely supervised whenever they access the Internet.

Back Cover Material

ENERGY FOR THE FUTURE • GLOBAL WARMING

Each book in ENERGY FOR THE FUTURE AND GLOBAL WARMING explores key concepts and issues related to energy availability, production, and use—and explains the impact of energy use on global climate. The eight volumes feature easy-to-read text accessible to a wide range of students, vivid photos, and clear, easy-to-understand illustrations and diagrams. This series will serve as a comprehensive and indispensable resource for anyone interested in the energy prospects for our future needs.

Each book in ENERGY FOR THE FUTURE AND GLOBAL WARMING includes the following features:

- Reader-friendly text style encourages quick comprehension
- Correlated to the middle school science curriculum
- Developed in consultation with a science curriculum consultant
- Clear explanations of key energy concepts
- A wealth of diagrams and other detailed illustrations that enhance understanding
- A glossary of scientific terms
- Key words defined within chapters

- Sources of further information, Web sites, and index

WIND POWER

Can you name a free, clean, and renewable energy source? The answer is blowing in the wind. In **Wind Power,** readers will discover how the historic windmill has become today's wind turbine and learn about the prospects for wind power in the future.

Titles in the ENERGY FOR THE FUTURE AND GLOBAL WARMING series:

- BIOFUELS
- FOSSIL FUELS
- GEOTHERMAL ENERGY
- HYDROGEN FUEL
- NUCLEAR ENERGY
- SOLAR POWER
- WATER POWER
- WIND POWER

Books For ALL Kinds of Readers

At ReadHowYouWant we understand that one size does not fit all types of readers. Our innovative, patent pending technology allows us to design new formats to make reading easier and more enjoyable for you. This helps improve your speed of reading and your comprehension. Our EasyRead printed books have been optimized to improve word recognition, ease eye tracking by adjusting word and line spacing as well as minimizing hyphenation. Our EasyRead SuperLarge editions have been developed to make reading easier and more accessible for vision-impaired readers. We offer Braille and DAISY formats of our books and all popular E-Book formats.

Made in the USA
Coppell, TX
05 November 2026

11292667R00033